P9-BYV-126

SIMPLE DEVICES

# THE PULLEY

Patricia Armentrout

The Rourke Press, Inc.
Vero Beach, Florida 32964

Patricia Armentrout specializes in nonfiction writing and has had several book series published for primary schools. She resides in Cincinnati with her husband and two children.

PHOTO CREDITS:
© Armentrout-Cover, pages 9, 16, 18; © East Coast Studios-pages 12, 13; © James P. Rowan-pages 6, 19; © Scott Barrow/Intl Stock: page 4; © Chuck Mason/Intl Stock: page 7; © Peter Langone/Intl Stock; page 10; © L.Rhodes/Intl Stock: page 15; © Frank Grant/Intl Stock: page 21; © Roger Markham Smith/Intl Stock: page 22;

EDITORIAL SERVICES:
Penworthy Learning Systems

**Library of Congress Cataloging-in-Publication Data**

Armentrout, Patricia, 1960-
The pulley / Patricia Armentrout.
p. cm. — (Simple Devices)
Includes index
Summary: Text and pictures introduce the pulley, a simple device used primarily to lift heavy objects.
ISBN 1-57103-178-2
1. Pulleys—Juvenile literature. [1. Pulleys.]
I. Title II. Series: Armentrout, Patricia, 1960- Simple Devices.
TJ1103.A76 1997
621.8'11—dc21 97–15148
CIP
AC

**Printed in the USA**

# TABLE OF CONTENTS

# DEVICES

**Devices** (deh VYS ez)—what are they? Are they trucks, cranes, and power tools? Are they wheels, **pulleys** (PUL eez), and ramps called inclined planes? Yes, these are all devices. Some of them are **complex** (KAHM pleks) and others are simple devices.

Complex devices, like trucks and cranes, have many parts. Simple devices, like wheels and pulleys, have few parts. This book looks at pulleys and where they are used.

*This heavy-duty pulley is used to lift objects that weigh tons.*

# THE PULLEY

Pulleys are simple devices used to lift heavy objects. A pulley is simply a wheel with a rope or cable running over it. The wheel has a groove, or track, in the center where the rope lies. The track keeps the rope from slipping off the wheel.

*This fixed pulley stays in place while the rope moves easily through the sheave.*

***Pulleys help construction workers move heavy steel beams.***

The grooved wheel, called a **sheave** (SHEEV), turns on an **axle** (AK sul). When the rope or cable is pulled, the axle allows the wheel to turn smoothly.

# A FIXED PULLEY

A fixed pulley is one that is attached to a fixed object, like a bar or beam. A flag pole has a fixed pulley at the top of the pole. The rope runs from the top to the bottom of the pole. The flag is attached to the rope ends to form a loop.

Pulling down on one side of the rope causes the flag to go up. This action is called change in direction. A fixed pulley gives the advantage of change in direction.

*Ski lifts use giant pulleys to move the chairs up and down the slopes.*

BRIKO

## CHANGE IN DIRECTION

Would you rather pick up a 50 pound weight, or use a fixed pulley to lift the weight? Lifting the weight takes a 50 pound force either way you lift it, but the pulley makes it easier because of change in direction.

A fixed pulley easily changes a downward pull on one end of the rope to an upward pull on the other end of the rope. Fixed pulleys are also used to raise window blinds and sails on boats.

*A pulley makes it possible to raise a sail without climbing the mast.*

# A MOVEABLE PULLEY

Moveable pulleys are often joined with fixed pulleys. Try this experiment.

Tie one end of a long rope to an overhead bar, or rod. Put the other end of the rope through the handle of a bucket, and pass the rope over the bar.

*It takes a 10-pound pull to lift a 10-pound weight using a single fixed pulley.*

*It takes a 5-pound pull to lift a 10-pound weight when you use a fixed and a moveable pulley.*

Your moveable pulley is now ready. When you pull down on the free end of the rope, the bucket will rise. The handle of the bucket is the moveable pulley that gives the **mechanical advantage** (mi KAN eh kul  ad VAN tij).

# MECHANICAL ADVANTAGE

Let's compare the fixed and moveable pulley to understand mechanical advantage.

With a fixed pulley, you would pull down 2 feet of rope to lift the load 2 feet off the floor. It takes a 50 pound pull to lift a 50 pound weight—there is no advantage here.

With a moveable pulley, you would pull down 2 feet of the rope to lift the load 1 foot off the floor. You apply less force over a greater distance. The moveable pulley allows you to use less effort to lift the load.

*Can you imagine lifting this heavy load without a pulley?*

# FORCE X DISTANCE = WORK

Cleaning your bedroom may be the way you define work, but scientists define work a different way.

Scientists measure how much force is used to move an object and how far the object is moved. Then scientists multiply these numbers to figure out how much work has been done. The formula is Force x Distance = Work. When you use pulleys, you can either apply a lot of force over a short distance or a little force over a greater distance.

*These boys use a simple pulley to make their work easier.*

# BLOCK AND TACKLE

Big devices like cranes rely on powerful engines to lift heavy loads. But some loads are so heavy that even the most powerful engines cannot lift them without the help of pulleys.

*Some work would be impossible to do without pulleys.*

*Pulleys on this crane are used to lower the seaplane to the water.*

When several fixed and moveable pulleys are combined, it is called a block and tackle. A block and tackle makes it possible for cranes to lift loads that weigh tons.

# PULLEYS AT WORK

A simple device like a pulley can be very useful. Pulleys are used in all kinds of equipment, like tow trucks, cranes, and oil derricks. Without the mechanical advantage of the pulley, these devices could not lift the tremendous loads they do.

Pulleys are found everywhere. Elevators use pulleys to move from floor to floor. Farmers use pulleys to lift hay bails to the top of the loft. Where do you see pulleys at work?

*Some pulley systems are complex but make work easier.*

Husky/
Bow Valley
VSBC-4

77594
38
38
77594
US

# GLOSSARY

**axle** (AK sul) — a bar that allows a wheel to turn around

**complex** (KAHM pleks) — made up of many parts or elements

**device** (deh VYS) — an object, such as a lever, pulley, or inclined plane, used to do one or more simple tasks

**mechanical advantage** (mi KAN eh kul ad VAN tij) — what you gain when a device allows you to do work with less effort

**pulleys** (PUL eez) — simple devices made up of grooved wheels that hold a rope or cable

**sheave** (SHEEV) — a wheel with a track or groove

*Large sailboats use many pulley combinations.*

# INDEX